RURAL
ARSON CONTROL STUDY

Prepared for
Federal Emergency Management Agency
United States Fire Administration

by
International Association of Fire Chiefs, Inc.
EMW-86-C-2080

This report has been prepared for the U.S. Fire Administration, Federal Emergency Management Agency (FEMA) under FEMA Cooperative Agreement number EMW-86-C-2080. All interpretations and opinions expressed are those of the authors and do not necessarily reflect the views of the government or the International Association of Fire Chiefs, Inc.

RURAL
Arson Control

Table of Contents

Arson Awareness Among Firefighters................................... 1
Cause Determination and Fire Suppression........................... 2
Fire Officers in Arson Control.. 3
Fire Scene Photography by Fire Departments........................ 4
Control of Arson Scene.. 5
NFIRS Reporting Accuracy... 7
Firefighter Training.. 8
Investigative Staff Workload... 9
Investigative Skills... 11
Regional Arson Units.. 13
Liaison with Law Enforcement Agencies............................ 15
Improve Accuracy in Uniform Crime Reports 16
"Total Burn" Syndrome 17
Prosecution/Courts 18
Juvenile Firesetter Counselling.. 19
Insurance Industry Support... 20
Insurance Industry Cooperation....................................... 21
Case Feedback 22
Inter-Agency Communication 23
Combat Public Apathy Toward Arson................................ 24
Public Education/Awareness.. 24

Introduction

Since 1978, the U.S. Fire Administration has been the focal point for Federal arson control programs. In this capacity, the USFA has provided technical assistance, resource materials, and training programs to state and local authorities. As part of this ongoing effort, the USFA and the International Association of Fire Chiefs entered into a cooperative agreement to evaluate rural arson control needs and to develop strategies for responding to them.

Project Approach

To develop a clear picture of the specific requirements of the rural arson control system, the IAFC conducted in-depth case studies of these systems in four rural counties. The results from visits to rural arson control programs in seven states are incorporated in this report. This project involved extensive interviews with fire, law-enforcement, and prosecutors. The culmination of the case studies in four communities was a "nominal group process" meeting at which elements of the community arson control system were represented. Attendees were asked to identify and rank specific needs of their own organizations. These needs assessments were discussed by the group and, by the end of the meeting, were prioritized.

IAFC staff reviewed the raw consensus statements to merge or eliminate duplicates from the four meetings. The twenty-two statements of need discussed in this report represent the essence of nearly fifty original need statements.

Arson control experts with special experience in rural arson problems were then selected to review these statements. We also asked that they evaluate and comment on all of the needs statements in a draft of this report. The objective was to elicit ideas about the validity of the statements, comments on possible solution options, and speak to other needs not previously identified.

Project Summary

Project findings can be summarized as follows:

- Rural arson control efforts are hindered by constraints and requirements specific to rural areas. These special conditions make rural arson prevention, detection, and control different from and more difficult than urban arson problems.

- Many challenges to the capability of rural fire departments to cope with arson detection and control are, in reality, connected to much more deep seated problems of the fire service in general and the volunteer fire service in particular. For example, many rural fire departments are not operating with the best information about rural arson control strategies and techniques. But the same is true for other topics of concern to the rural fire service. Arson prevention and control is another instance of the challenge facing fire departments trying to stay abreast of a wide range of responsibilities.

- To improve arson control in rural areas, special initiatives across a wide range of issue areas are indicated. In particular, state agencies need to play a stronger role in assisting rural communities manage their arson prevention and control programs.

- The United States Fire Administration both directly through its anti-arson efforts and indirectly through its other programs can improve rural arson control.

- The United States Fire Administration can perhaps achieve the maximum leverage of its funds by packaging and disseminating specific field-tested solution strategies and alternatives. Rural fire departments and investigators tend to be isolated from the mainstream of information about effective arson control strategies. Special efforts will be needed to develop and maintain contacts with the grass roots level.

- A greater use of law enforcement, insurance industry, and community-based resources can and should be made. State agencies can play a more effective role in sponsoring cooperative efforts to control arson control problems facing rural areas. One notable example is the development of state-sponsored arson strike forces that can respond on call to a serious arson problems in rural areas.

Concentrated efforts to make increased use of resources that exist outside the fire service in both the public and private sector perhaps offer the greatest opportunities for achieving anti-arson objectives.

The general and more specific findings that follow contain no startling revelations. Nor do they call for massive governmental expenditures. Rather, they reinforce the findings and the common sense judgments of most current observers in this field that many improvements need to be made. Many of these improvements may be minor in themselves. Collectively, these improvements can reduce the pressures on rural resources and direct more pressure instead on rural arsonists. This report provides a framework for looking at how existing solution strategies could be more effectively harnessed.

Report Structure

This report consists of twenty-two mini-chapters, each of which is divided into four parts:

- A statement of need;
- A discussion of the statement of need;
- A tally of the number of arson control experts polled who agreed or disagreed with the statement;

- A brief discussion of existing programs around the country that address all or part of this need.

Each mini-chapter deals with one topic, corresponding to a priority need identified by to topics identified by By breaking down rural arson needs in this format, we hope readers will find it easier to first review topics of greatest interest to them.

Arson Awareness Among Firefighters

Need Statement:

Need to raise the level of awareness among firefighters about their role in arson prevention, detection, and investigation.

Discussion:

Some respondents noted that rural firefighters do not rank arson as a high priority. Other subjects, such as hazardous materials, are a much greater concern. In some communities the desire was expressed for a training program that would orient firefighters to the importance of their contribution to arson control.

Three of the four counties had recent experience with the problem of firefighters themselves becoming fire setters and felt that such a training program could help to deter this behavior.

■ Arson Control Expert Concurrence

Agree 23   Disagree 1

Solution Options:

Develop either a training program or program components (such as audio-visual materials), designed specifically for the needs of the volunteer fire service, that could be integrated into the regular training schedule. Such materials would have to be divided into blocks of one or two hours each to accommodate the normal training pattern of volunteer departments.

Another technique is to return initial crews to the scene to view findings so they become aware of procedures used by investigators. This usually can be done "in service."

Put on an arson awareness program like the one the York County (Pa.) Fire Chiefs and Firefighters conducted. It consisted of 10 one-hour slide presentations to break it down into manageable time blocks and reinforce its message over a longer timeframe.

Spotlight local fire and arson investigation experience in locally-delivered arson awareness courses for firefighters.

Enact local or state legislation that requires investigators to be summoned when cause cannot be definitely established as accidental.

One community highlights arson awareness in its annual report by specifically enumerating the responsibilities of the fire chief. Determining fire cause is specifically given as one of these responsibilities.

Cause Determination and Fire Suppression

Need Statement:

Initial cause determination should be an integral part of fire suppression.

Discussion:

It is by no means a universally accepted tenet among rural firedepartments that initial cause determination is one of their prime responsibilities. Many progressive departments recognize the importance of cause determination to arson control, fire prevention, and public education. Some rural departments still define their role strictly as fire suppression. Because of the diversity of positions on this point, rural fire departments observed during the course of this study ranged from departments with specially trained and highly motivated investigative teams to departments with no interest in even performing the most perfunctory cause determination.

To encourage rural fire departments to take as active a role in cause determination as local resources allow and local conditions require will take additional efforts at the national and state level.

■ Arson Control Expert Concurrence

Agree 20   Disagree 4

Solution Options:

Develop a set of model procedures that fire departments could adopt to guide their actionsat a fire scene. For example, these might include the proper method for roping off a scene, rules for control of smoking materials on the fire ground, or recording the license numbers of spectators' vehicles;

Require reporting of all fires and causes to a central state agency, such as the state fire marshal's office. Kansas uses this method, and it has improved the quality of reporting. In turn, cause determinations have increased. Use of "fire-police" for crowd control and protection of the scene.

By Pennsylvania law, fire chiefs must sign certifications of acceptanceof responsibility for fire investigation. Clearly mandating fire departments to discharge this responsibility provides a formal basis for this activity. Other steps (including as a last resort, a range of sanctions) should also be available to state authorities to ensure that this responsibility is carried out. Formal mechanisms that spell out responsibilities need to be augmented by a variety of reinforcements. Reinforcements could include the appointment of a county fire marshal or administrator to monitor this responsibility, making cause determination a part of every training program for promotional qualification, and making reporting of suspicious fires to the appropriate enforcement agency mandatory.

Each department should maintain constant round of training programs to reinforce the message. This may require utilizing assistance from state agencies, colleges, manufacturers of fire detection apparatus, insurance companies, etc. for materials and other resources. Two related suggestions are:

- Reprint the "Arson Control Guide for Volunteer Fire Departments" (FA-51 January 1981). It is an excellent resource document that deserves far-wider dissemination. Reprinting thisguide (and perhaps preparing alow-cost pamphlet to advertise its features) appears warranted.

- Encourage state fire marshal offices to distribute the "Arson Control Guide for Volunteer Fire Departments" to newly-appointed/elected fire chiefs in rural areas.

Fire Officers in Arson Control

Need Statement:

Upgrade the capability of rural fire officers to contribute to the arson control process.

Discussion:

Fire officers of rural departments would be more motivated to contribute to arson control if they were given training and some basic reference materials.

Rural departments must rely on a number of personnel to perform the chief officer role at the fireground. Many of these officers may not have received sufficient training nor have had enough experience to correctly judge the cause of a fire.

Officers in rural departments need to be motivated if they are to be expected to carry out sound cause determination practices. Many officers remarked that their motivation would increase if the conviction rate for arson cases increased. Others pointed to the disincentive of having to lose time from their primary jobs to testify in court.

■ Arson Control Expert Concurrence

Agree 23   Disagree 1

Solution Options:

Minnesota Statute 299F.051 provides for $35.00 per day plus expenses to volunteers for time spent in attending fire investigation training courses, when they attend with a law enforcement officer from their community.

Exemplary performance certificates. State agencies could contribute to improved cause determination performance by recognizing outstanding contributions of fire officers to arson case development. Recognition could range from a departmental letter of commendation to a monetary award.

Direct feedback from investigators may, however, be all the motivation that many fire officers need. Investigators can recognize individual contributions with a phone call and thank department-wide efforts at monthly firefighter's meetings. As simple as these two feedback mechanism are, they are proven incentives to increase arson detection awareness among fire officers.

Arson control awareness can be promoted through other avenues. It needs to be soundly based in state statutes. Therefore state laws should clearly assign the responsibility for fire cause determination and reporting. Fire officers might be provided with a pocket guide or outline of their responsibilities in cause determination and reporting.

States and/or locales that establish qualification requirements for fire officers should include fire cause determination proficiency as a critical area.

Fire Scene Photography by Fire Departments

Need Statement:

Fire departments should have the capability to photograph or videotape the incident, fire growth, suspected area of origin, and any evidence found at the scene.

Discussion:

Rural fire department personnel are in a position to greatly assist investigators by gathering photographic evidence at a fire scene. Some departments carry their camera equipment on apparatus and use it as soon as personnel can be spared from fire fighting duties. Pump operators can often take pictures of the exterior and crowd between other responsibilities. Initiation of evidence gathering by firefighters and fire officers can, in part, compensate for non-availability of an investigator (due to staff shortage, distance problems, etc.). This would at a minimum require:

- availability of appropriate (low cost, easy to use and maintain)equipment;
- training in proper fire scene photography techniques;
- integration of this practice into the investigative process; and
- assigning the responsibility to one or more persons to ensure this capability is available (often a department has a "camera buff" willing to take on this responsibility).

■ Arson Control Expert Concurrence

Agree 24   Disagree 0

Solution Options:

Low cost compact cameras designed for underwater use have been used in fire incidents with some success. They are rugged, waterproof, designed for use by someone wearing gloves, and inexpensively priced. (many retail for less than $200)

Several jurisdictions use videotape equipment to record fire growth. In other areas, one photographer takes fire scene pictures for a number of departments which share the cost for this service.

Rural departments might be aided by a list of cameras that other fire departments have found suitable (Rehobeth Beach, Delaware uses a Minoltatm 35mm. camera that is both shock and water resistant).

State or federal agencies could negotiate guaranteed purchase price contracts with vendors for several types of field proven cameras/video camera recorders. This service would help departments obtain the suitable equipment for an affordable price. This would encourage the practice of the officer-in-charge photographing the fire scene.

Prepare a videotape for the volunteer fire service describing the benefits of fire scene photography and videotaping. The videotape could outline practical tips for enhancing fire scene documentation.

Control of Arson Scene

Need Statement:

Strengthening is needed in rights and procedures for scene control, chain of custody, and searches.

Discussion

The effect of the U.S. Supreme Court decision: Michigan v. Clifford has been to complicate cause and origin investigation. This ruling reinforces the need to obtain the owner's written consent (stating that the consent was given voluntarily, without coercion, and that the individual understood that it did not have to be granted) for conducting an arson investigation.

A second action that needs to be taken is to begin the investigation before the engines leave. Warrantless re-entry after control of the fire scene is relinquished is discouraged by this decision. In the absence of some reason related to fire safety or the owner's written consent, an administrative or court-approved search warrant is needed to investigate the cause of the fire once fire suppression and control activities cease. The administrative warrant secured to determine cause should not be construed as permitting a wholesale search of the property. Once the cause is determined, the administrative search should end and any broader investigation should be done after a criminal search warrant has been obtained. This means interrupting the search, securing the scene, swearing out an affidavit, and obtaining a criminal warrant.

The practical implications of this new ruling will make it more difficult for both fire officers and investigators to complete a fire scene investigation..

Michigan v. Clifford illustrates the increasing need to keep fire officialsabreast of legal developments that affect fire cause practices. The decision further pointsout the need for close working relations with local prosecutors. Local law enforcement authorities may also be able to offer sound advice in proper search procedures.

In rural areas obtaining a search warrant can mean delays of several hours before investigation of the scene can begin. It often involves a round trip drive of many miles for the investigator to return with a search warrant. Unless arrangements are made to post guards to secure the property it can lead to a loss of custody of the scene.

■ Arson Control Expert Concurrence

Agree 23   Disagree 1

Solution Options:

Efforts should be made through the U.S. Fire Administration and state fire service organizations to publicize this and other landmark decisions.

Owner/occupant consent forms, and, if appropriate, administrative search warrant forms should be carried in response vehicles.

Develop procedures for rapidly securing a search warrant. States that grant fire marshals administrative search warrant authority save their rural investigators valuable time and energy

Klamath County, Oregon's Arson Strike Force has developed procedures for control of scene as follows:

-       When members of an on-scene fire service encounter a fire of a suspicious nature (the Strike Force does some arson recognition training), they alert the Strike Force through their dispatch center.

-       Dispatch alerts four members (cause determination investigators) of the Strike Force's investigative unit.

-       Firefighters maintain control of the scene until an investigator arrives but <u>do not</u> proceed with investigation.

The organizers of a pilot arson strike force in Northeast Oklahoma have found that adding an attorney to the investigative team is useful. Prosecuting attorneys can advise investigators on technicalities associated with evidence collection and custody, suspect interviews, subpoenas, consent to search forms, etc.

NFIRS Reporting Accuracy

Need Statement:

There is a need to improve fire cause data reporting on National Fire Incident Reporting System (NFIRS) forms.

Discussion:

In none of the jurisdictions surveyed was there observed a systematic method for updating NFIRS reporting forms. At present, a fire officer, upon alerting a state investigator, typically designates the fire cause as either undetermined or suspicious. The fire department then forwards an initial fire reporting form to the appropriate collection agency before obtaining an update from the investigator. The state fire investigator's determination of cause may never reach the stateagency responsible for maintaining fire incident reporting. Even when the investigation is handled entirely by local investigators, the same problem can arise- the investigator's final cause determination not being incorporated in the fire incident report.

Many fires that are currently reported as undetermined/unknown have, in fact, been determined, just not reported. As a result suspicious and incendiary fires are likely to be seriously under-reported in rural areas. This, in turn, can effect the amount of resources allocated at all levels of government to rural arson control.

A related problem arises from fire officers not understanding the underlying definitions of fire cause used in the NFIRS reporting system. For instance, several rural departments routinely reported juvenile arson as 'children playing with matches". NFIRS coding instructions need to clarify that intentional fire setting by children using matches should be classified as suspicious or incendiary. The cause of "children playing with matches' should be restricted to those incidents in which children inadvertently set fire to other materials in the course of playing with matches.

■ Arson Control Expert Concurrence

Agree 24   Disagree 0

Solution Options:

Revise NFIRS materials to ensure that guidelines are suggested to strengthen reporting practices (This might be accomplished through the National Fire Information Council.). Investigators should complete a revised 902 form upon completing the fire investigation and submit copies to both the originating fire department and the state NFIRS unit.

-       Maryland's Fire Marshal's Office has an instructor on staff (actually, a staff person assigned to instruct) whose role is to teach firefighters the proper method for filling out Fire Incident Reporting System forms.

-       One proposal for simplifying NFIRS reporting on fire cause is to substitute the following set of questions for the present coding formats:

- Has a cause for the fire been determined? __ Yes __ No
- If yes, state the cause _____or cause code __.
- If no cause has been determined, what is the current status?_____
  This change merits consideration by the appropriate NFPA 901 committee and by the NFIC.

Firefighter Training

Need Statement:

Options for firefighter training that will maximize/optimize their effective contribution to arson control.

Discussion:

All of the jurisdictions/agencies contacted referred to improved firefighter training as a factor that could significantly contribute to arson control. All ranked it high. Recent arson control studies suggest that up to three quarters of all firefighters have never received arson control training. A relatively high turnover rate for firefighters is a chronic problem in many volunteer fire departments. Maintaining arson detection skillsamong "back step" firefighters will likely prove a recalcitrant problem. Making such training a part of basic training appears to be the surest means to ensure this training is received.

Topic areas suggested for this training included:

- arson recognition
- preservation of evidence (when and how to secure evidence at the scene and how to meet chain of custody requirements when the evidence has to be moved)
- motivating firefighters to be concerned with fire cause

■ Arson Control Expert Concurrence

Agree 24   Disagree 0

Solution Options:

New York State, Department of State, Office of Fire Prevention and Control offers a training program, "Fire behavior and Arson Awareness," directed at line firefighters. The course is 12 hours long and is given in the firefighter's own community. About 12,000 firefighters have been trained since 1979. The Indiana State Fire Marshal's office has presented a 12-hour "cause and origin" arson detection class to about 7,000 of the state's firefighters over the past 10 years. This class applies against a training requirement of at least seven hours per year.

The Ontario (Canada) Provincial Fire Marshal's Office has recently embarked on a training program for rural firefighters. The program is based on one that has been conducted for urban firefighters for more than a decade; however, it has been re-formatted for presentation on two consecutive weekends.

Oklahoma, <u>which pays for fire service training</u>, has developed arson-related programs for firefighters. Special attention has been given to the needs of volunteer firefighters by:

- Training of firefighters in a 12 hour arson recognition course (primarily for volunteers). This has apparently resulted in an increase in convictions out of rural departments;
- Making use of existing videotape capability;
- Redesigning courses so that they can be taught on weekends and in the evening.

Investigative Staff Workload

Need Statement:

Rural jurisdictions, especially in states where state agencies (Fire Marshal, State Police) supply investigators, are subject to staff shortages that can adversely affect case management capacity.

Discussion:

This was identified as a major problemboth through interviews and through the "nominal group process" workshops. Three of the four case study jurisdictions cite one or more problems arising from a shortage of trained arson investigators. This may lead to:
- "windshield investigations" perfunctory investigations conducted by investigators who barely have the time to leave their vehicles because of their case backlog
- slow response time by investigators;
- large portion of investigator's time spent in transit when he/she must cover a wide area;
- response conflicts when more than one request is Rending ;
- shortage of time for liaison with firefighters;
- incorrect or inadequate reporting of fire statistics;
- role conflict (where investigators also do inspections and certifications).
- lack of backup staff, etc. cause chronic overextension of state investigative resources

Historically, state fire investigative resources in most states have never been staffed at a level needed to respond to all requests for assistance in a timely manner.

Distances that must be travelled also pose a problem. Even in a small state with a relatively large number of investigators relative to size of population, arrival of a trained investigator frequently takes one to one-and-one-half hours.

Additional duties imposed on investigators can seriously eat into time available for investigative activities. For example, inspecting nursing homes is a revenue producer for state governments. Accordingly, state fire marshals who also perform double duty as fire investigators and inspectors frequently encounter pressure to meet their nursing home inspection quota.

No states reviewed as part of this study have developed guidelines to establish whether or not investigative workloads are excessive. This is understandable in light of the fact few state officials expected additional investigative resources to be authorized by state legislators. Absent a sharp drop in arson incident rates, or local investigative resources assuming greater responsibility for investigating rural arson, state investigators were seen as likely to remain chronically over-extended..

■ Arson Control Expert Concurrence

Agree 24   Disagree 0

Solution Options:

One or more of the following options might help statesand local resources meet their common responsibilities, including:.
-    Establish alternative working relationships such as having local law enforcement agencies assume a greater role in investigative activities;
-    Assign investigators to full-time investigative duties;
-    Increase manpower through measures such as insurance premium surcharges;
-    Assign more investigators;
-    Reduce the number of investigations requiring state investigators to respond by making more efficient use of local fire officials to screen-out fires that should be handled locally
-    Manage caseloads by assessing case solvability factors before investing time in follow-up investigation. This would be a formal program involving investigators and their supervisors in a closely-monitored case management system

State and local officials concerned about investigators' workloads would benefit from a USFA-sponsored study that surveyed state investigative workloads, established criteria for determining unacceptable caseload levels, and laid out alternative strategies to reduce excessive investigator workloads.

New York State, Department of State, Office of Fire Prevention and Control has adopted an approach to arson investigation that emphasizes maximum development of local capability. This approach provides investigative training and services to local investigative teams. The result has been a substantial increase in the number of trained arson investigators available in rural areas.

Wayne County, New York, which has a team trained under the above system, uses the following approach:

When a local Fire Chief calls out the investigating team, two fire volunteers and one law enforcement investigator respond. The county has an evidence collection van but investigative team members carry evidence collection kits in their cars. The Wayne County, New York program uses five volunteer investigators from fire services (Wayne County has several volunteer fire departments) and three from the Sheriff's Department (force size is about 20 Deputies). Fire service investigators are responsible for taking samples and photographs; sheriff's deputies are responsible for interviewing and report writing. State Policealso participate in some parts of the county (where they have substations), replacing Sheriff's Deputies.

The arson task force concept remains one of the best mechanisms to overcome inter-agency communication barriers. Similar but less formal organizational approaches have produced improved inter-agency cooperation and coordination. The Southern Ohio Fire Investigators Association and the Santa Barbara (California) County Arson Investigator's Association are two examples of professional interest groups that meet regularly around a training and intelligence sharing format.

The "lessons learned" by these and similar associations could be gathered and distilled to promote their successful development in other rural areas.

The Oklahoma Association of Arson Investigators has recently established an "Arson Task Force" that furnishes teams of experienced investigators to communities in the Northeast part of the state. Teams are drawn from a pool of volunteers and are provided at no cost to the requesting jurisdiction.

Klamath County, Oregon's "Arson Strike Force" provides teams of experienced, working cause and origin personnel to determine cause and gather evidence at suspected arson incidents. These teams act as a screening mechanism for the Oregon State Police Arson Unit which is responsible for investigation of incendiary fires.

The Central Virginia Regional Arson Investigation Squad has 65 trained law enforcement officers available on call to respond to any arson incident in a two county, 10 jurisdiction area.

The New Jersey Attorney-General's Office has assisted the formation of 16 county-wide arson investigative units. Rural counties lacking either the demand or resources are exploring the possibility of forming multi-jurisdictional units.

State firefighter's and fire chiefs' associations could provide state-wide leadership to obtain additional resources if motivated. In Delaware it was reported that such an effort resulted in the establishment of regional offices with more manpower assigned to perform investigations.

Investigative Skills

Need Statement:

Arson investigator trainingprograms need to stress skills development in areas other than fire cause determination.

Discussion:

Skills in conducting criminal investigations are of special importance to arson investigators in rural areas who must often work alone. No systematic national analysis of state arson investigator training requirements has ever been made.

Some of the areas where improved skills were sought by investigators contacted in the four study areas included:

- interrogation/interview   technique;
- report writing;
- evidence collection
- legal issues.

Investigators cited high caseloads, lack of agency support, and non-availability of local training sites as barriers to obtaining additional training in these areas.

■ Arson Control Expert Concurrence

Agree 24   Disagree 0

Solution Options:

At least one state requires arson investigators to complete forty Continuing Education Units every three years.

Conduct a national assessment of investigative requirements for state arson investigators that would lead to recommendations that would help local and state agencies review their own requirements and determine if additional training requirements were warranted.

Investigative Equipment

Need Statement:

Investigator performance is directly affected by the availability of appropriate investigative equipment and supplies.

Discussion:

In two of the four sites surveyed, the state investigators were not provided with even basic investigative equipment. Some examples of basic equipment shortages included:

- hand-me-down cars prone to breaking down
- cars too small to transport basic investigative equipment
- state agencies not issuing even simple hand tools (shovels, rakes, sifting gear), turn-out protective equipment, and
- specialized evidence storage bags, cans etc.

More sophisticated devices like hydrocarbon detectors were simply not provided to individual investigators. State investigators pointed out the need for portable generators to supply lighting on scene and hand-held dictation equip ment so investigators could complete reports while travelling between investigations.

The direct impact of this lack of support for investigators is to hinder them in the normal performance of their duties. Productivity is needlessly sacrificed when investigators are not equipped with basic investigative and administrative supplies. Indirectly, it undermines morale. Many investigators interviewed interpreted the lack of equipment as symbolic of the lack of interest on the part of their elected and appointed leaders in the arson control mission.

Current literature in the field of arson investigation reveals a number of suggested basic equipment and supplies listings. State and local officials might find it easier to justify expenditures to meet a minimum equipment listing if one nationally-recognized standard were adopted and publicized.

■ Arson Control Expert Concurrence

Agree 24   Disagree 0

Solution Options:

State and local investigative agencies should be made aware of the Anti-Arson Funding Guidelines pamphlet produced by the insurance industry.

Agencies unable to secure funding though normal budgetary channels might find local and state insurance organizations willing to grant funds to purchase investigative equipment designated as a national standard.

Regional Arson Units

Need Statement:

Regional program for focusing arson control resources in areas with a high arson rate.

Discussion:

Multi-agency arson investigative units on the national, state, and regional level have all amply demonstrated their effectiveness in controlling outbreaks of arson. One barrier to their more widespread use is lack of publicity on and details about such organizations are formed and operated., Such a unit would bring to bear local, state, and perhaps Federal and private resources for a finite period to help local officials attack an arson problem beyond their resources to control.

■ Arson Control Expert Concurrence

Agree 22   Disagree 1   No Comment 1

Solution Options:

Ventura County, California and Memphis-Shelby County, Tennessee, among others, have established county-wide units. The Regional Arson Investigation Squad of Central Virginia is an example of a multi-county program as is South Nevada Regional Arson Task Force. The Michigan State Police have established a state-wide Strike Force program. In California, the Governor's Special Arson Task Force has formed an Operational Group to provide on-site investigative assistance on the request of any local agency. Both these state and regional programs have proven successful in assisting local jurisdictions.

The brief description that follows is of two such regional units. They are representative of the way a number of regional programs are organized and operated.

The Klamath County, Oregon Arson Strike Force (ASF) operates under a formal charter with oversight from the County's Arson Task Force (an advisory group composed of fire chiefs, police chiefs, county council members, etc.).

The ASF has one representative from each participating fire service and law-enforcement agency in the County. These are the persons who have assigned responsibility for arson detection/investigation within those agencies; they are field investigators rather than managers.

Apart from oversight by the Klamath County Arson Task Force, the Strike/Force runs itself. They conduct monthly meetings that incorporate training and review or update of information from previous sessions.

The ASF currently has approximately 15 members about half of whom are experienced fire cause investigators. Insurance company participation is encouraged but is limited to an advisory capacity and to provision of training and equipment. (An active insurance company role might be seen as a conflict of interest.)

The Strike Force uses both paid and non-paid, fire and police personnel. Participation is voluntary, but the individual participant must pass evaluation by membership of the Strike Force. Participation, as stated above, is limited to those within the agency who are already responsible for fire causeinvestigation.

The ASF is an extension of inter-agency mutual aid agreements that existed prior to its formation. This is noteworthy since it is simply a new use of an established concept. As such, it already had a regulatory basis for its existence. Thus, it is possible that this approach could be readily applied in other parts of the country within the scope of existing authority bases.

In Northeastern Oklahoma, a regional arson program was implemented in the late spring of 1984. Referred to as an "Arson Task Force", the program makes use of strike force type teams to assist arson investigators in Northeastern Oklahoma. The program is strictly volunteer with teams composed of firefighters, police, media representatives, and an attorney. Currently, the task force has enough volunteers for four or five teams.

Assistance is provided at no cost to the receiving jurisdiction. The program requests but does not demand reimbursement for such things as film and evidence cans. Planners are currently assessing a funding scheme that would be based on insurance company subscriptions.

The program currently has a van with equipment and is trying to arrange another. They also have been able to obtain some complimentary investigative equipment from manufacturers who want to have their products displayed and/or tested in the field. One fire dept provides support for the van and supplies the driver (in-kind donation).

Another program approach that has been implemented successfully is the fire prevention cooperative. The concept entails formation of a joint business/ government organization made up of representatives of firms and agencies concerned with fire prevention in the target area. It is a multi-level, inter-jurisdictional approach that emphasizes increased public awareness as a means of preventing man-caused fires.

The first such program we are aware of is the Rogue Valley, Oregon, Fire Prevention Cooperative. Formed in 1976, the cooperative has served as a model for similar programs in the Pacific Northwest and elsewhere. There are currently 16 such coops in the northwest (principally in forested areas) and two in Oklahoma. Similar programs have been attempted in Arkansas and Alabama with limited success.

All of the coops encountered thus far have been in forested areas where wild-land fires area particular problem. In practice, they are used to reduce both wild-land and structural man-caused fires, whether incendiary or accidental.

Liaison With Law Enforcement Agencies

Need Statement:

Arson investigators need to cultivate and maintaingood working relations with other area law enforcement personnel.

Discussion:

In rural areas in particular, arson investigators must rely on the resources of other agencies for support services (crime scene technician, polygraphs, access to professional journals, legal decisions, intelligencesources, etc.). This requires that they develop and maintain good relations with agencies that have these resources.

By maintaining effective working relationships with law enforcement agencies in the area, fire investigators can have access to this information at the least cost to the public. Having to rely on another agency's resources may have the beneficial effect of bringing about better working relationships. For example, in two instances state investigators figuratively camped out in another agency's office space (in one case a state police barracks, in another a fire departments headquarters.) In both instances, sharing office space strengthened the working relationships

■ **Arson Control Expert Concurrence**

Agree 24   Disagree 0

Solution Options:

Investigators consulted during field interviews suggested the following as helpful to closer working relationships with law enforcement resources:

- Crediting the "collar" to a police agency;
- Ensuring that the media spotlight other agencies that have contributed to a successful investigation;
- Extending many small courtesies (inviting crime scene technicians to arson investigator meetings,. sending Christmas cards, buying donuts etc.);
- Receiving training outside ones own discipline (fire classes for law enforcement officers and vice versa);
- Exchanging intelligence.

Improve Accuracy in Uniform Crime Reports

Need Statement:

Greater accuracy and uniformity is needed in the completion of Uniform Crime Report (UCR) forms.

Discussion:

Agencies often unknowingly mis-report arson data to the UCR system. Many factors contribute to this situation. Rural areas face special problems in complete and accurate reporting. Among those problems observed were state and local law enforcement agencies failing to work out agreed procedures for reporting arson incidents, volunteer fire departments failing to report incidents of arson to a state or local law enforcement agency, and local law enforcement agencies only tallying those arson incidents in which its own officer filed an initial complaint.

because arson is handled as a special reporting task with its own form, it is frequently assigned to personnel outside the normal UCR reporting system. These individuals may have none of the training, reference materials, or editing checks that those responsible for UCR reporting may have. If improved occurrence in UCR reporting of rural arson incidents is desired, additional efforts will be needed.

■ ,**Arson Control Expert Concurrence**

Agree 24   Disagree 0   No Comment 2

Solution Options:

Simplify UCR reporting practices by making property classifications identical with NFIRS property classes. Prepare special guidance for fire and arson reporting authorities to clear up misunderstandings in the use of 901 and UCR definitions and terminology.

Revise and expand UCR guidance materials to give specific guidance on how to correctly code problematical incidents.

Such guidance should stress the pay off to departments in accurately reporting findings. State agencies should consider developing incentives to improve reporting compliance. Ensure also that each reporting agency has been provided with a copy of all current guidance on the new, incident-based UCR reporting system

Provide training (including periodic refreshers) to those who fill out UCR forms.

Require certification for those completing UCR reports.

"Total Burn" Syndrome

Need Statement:

Determining origin and cause is frequently complicated in rural areas by the extent of destruction. This "total burn" syndrome places additional pressures on rural investigators to preform at peak professional levels with minimal support.

Discussion:

A rural arson investigator is frequently confronted with the task of culling evidence from the remains of a structure that is largely or totally destroyed. (Indeed, fire damage in rural areas runs at least three times urban rates, according to the best available information.) A single investigator with a heavy workload does not have the time or the physical capability to dig out a building when it is "on the ground." Both local and state investigators, therefore, need to have ready access to additional manpower and, sometimes, heavy equipment to thoroughly determine cause in this regard. investigators mentioned that small end-loaders, often referred to as "Bobcats," can prove particularly valuable in certain applications.

Investigators in two of the four sites stated that they encountered practical difficulties when they approached volunteer fire departments for manpower needed to dig down through the rubble to the expected area of origin. This often means that in the absence of some strong indication of arson the investigator, lacking the resources to move and examine debris, will perform only a perfunctory investigation.

Given that arson appears to be most frequent in economically depressed areas, one can justifiably ask how jurisdictions that are already money-short will be able to fund the necessary labor and equipment for competent investigations.

■ **Arson Control Expert Concurrence**

Agree 24   Disagree 0

Solution Options:

National Advisory Committee members suggested that investigators canvass local resources for heavy equipment needs. Sources cited included: county and town public works, emergency management agencies, utility companies, insurors, the construction industry and others in the private sector.

Make arrangements in advance with departments requesting investigation that if manpower is needed to help dig out the scene, it will be provided.

Prosecution/Courts

Need Statement:

An effective arson control program requires effective prosecution of arson cases.

Discussion:

There was general agreement among the study's four working groups that successful prosecution of arson cases is more difficult than for other major crimes. A variety of difficulties were cited to support this view including:

- cases must be proven twice, first to show that a crime has been committed, second to show the accused was/ were responsible;
- public apathy toward arson, particularly arson for profit in economically depressed areas;
- firefighter inexperience with investigative needs may lead to inadequate preservation of evidence or breaks in the chain of custody;
- the rapid staff turnover that as a rule exists in the prosecuting attorney's office makes it difficult to keep attorneys experienced in arson prosecution;
- complexity of cases and arson law;
- prosecutor inexperience with arson law;
- low probability of conviction; and
- uncompensated court appearance cause hardship for volunteer firefighters and affect their willingness to appear as witnesses.

Compounding these difficulties are the fact that prosecutors in rural areas may only serve on a part time basis and/or have heavy caseloads. This may leave them fewer opportunities for special training in arson prosecution.

■ **Arson Control Expert Concurrence**

Agree 24   Disagree 0

Solution Options:

The Florida State Fire Marshal's office has an attorney on staff whose function is to assist with preparation of arson cases and, if necessary, to prosecute them. Florida enacted a new, simplified arson law in 1979; this has increased prosecutors' willingness to take on arson cases.

Klamath County, Oregon's "Arson Strike Force" uses a team of four experienced cause and origin investigators at each suspected arson. This permits quick response and improves the (quality/reliability) of evidence. This has increases the confidence of prosecutors in the quality of cases they get.

The Northeastern Oklahoma arson strike force counts as a member an attorney experienced in arson prosecution. His participation improves the quality of case reports and other supporting documentation, thereby increasing the likelihood of effective prosecution.

In New Jersey the states' attorney general sought out cases that local prosecutors regarded as marginal. By successfully prosecuting the majority of these cases, the attorney general encouraged more aggressive local prosecution of subsequent arson cases.

One state has offered a three hour training course to local prosecutors. States attorney generals could offer special arson seminars to local prosecutors in conjunction with other professional meetings and training opportunities. As a minimum, the state's central prosecution office should seek to make arson prosecution expertise available to local prosecutors.

Prosecuting offices can assist arson investigators greatly if they designate one or more prosecutors to handle all arson cases and respond on scene upon request.

Prosecutors should advise the court that volunteers will be called as witnesses and, if at all possible, there should be an avoidance of delay for those individuals that are uncompensated.

Juvenile Firesetter Counselling

Need Statement:

Counselling of juvenile fire setters in rural areas.

Discussion:

Rural areas face particular problems in providing counselling to juvenile fire setters. In a many rural counties, demand for service is not likely to warrant setting up separate counselling programs. Regional counselling programs offer a possible alternative. Both county and regional counselling programs will have to rely on fire officials and arson investigators to identify those candidates most in need of assistance and those families most likely to benefit from counselling programs. Even the smallest volunteer department plays a front line role in identifying firesetters and referring them to appropriate authorities.

■ **Arson Control Expert Concurrence**

Agree 24   Disagree 0

Solution Options:

Provide distributionof the U.S. Fire Administration's, Juvenile Firesetter Counselling Guides to all rural departments.

The Missoula, Montana Rural Fire District has recently begun implementation of a rural version of San Francisco, California's "Fire Hawk" program for juvenile fire setters.

Regional mental health centers may, if properly approached and utilized, prove an effective resource.

Information could be gathered on examples of successful counselling programs for juvenile fire setters and other alternative efforts (community service, restitution, arbitration panels, etc.).

Use of the "Learn Not To Burn" program in schools in Southeastern Oklahoma has shown success.

Insurance Industry Support

Need Statement:

Active support from the insurance industry would enhance rural arson control efforts.

Discussion:

Insurance industry support could enhance local arson control efforts through their financial resources.

In general, experience indicates that local insurance organizations are receptive to requests for funding, especially when the request is for equipment and the proposal is well documented and presented.

This support appears to be more forthcoming when insurers, adjusters, and others are represented in local arson task forces or arson investigation associations.

■ **Arson Control Expert Concurrence**

Agree 24   Disagree 0

Solution Options:

The Northeastern Oklahoma Arson Strike Force is currently exploring options for insurance industry donations to provide investigative equipment such as hydrocarbon detectors as well as for consumables (film, evidence cans, etc.) used in strike force investigations.

Fire investigators and fire departments should be made aware of the program guidelines published by the insurance industry that recommend criteria for local and state insurance associations consider in making anti-arson grants.

The U.S. Fire Administration might include a special section in future editions of the Arson Resource Directory on securing insurance industry grants. Further, the Arson Resource Center could maintain a supply of the anti-arson grant guidelines

Insurance Industry Cooperation

Need Statement:

Improve cooperation between fire service and insurance industry arson control efforts.

Discussion:

There is a general perception among the sites studied that insurance industry practices contribute to the arson problem in rural areas by over-insuring properties, and are too quick to pay off claims even when they occur under suspicious circumstances. Fire service, law enforcement, and state fire marshal personnel would benefit from better communication. As a general rule there appeared to be scant regard for the problems of other actors in the systems. For instance, fire service personnel maintained that in many cases properties had highly inflated valuations placed on them. Insurersdid not attempt to assess the values claimed by the insured. These critics wanted insurors to inspect the property or otherwise satisfy themselves as to the value of the property prior to underwriting them. Insurors, on the other hand, pointed out that the time required to check up on the values claimed would increase costs and cut into their selling opportunities.

■ Arson Control Expert Concurrence

Agree 23   Disagree 1

Solution Options:

Conduct a study to document programs that have improved working relationships among the insurance industry and rural arson investigators. The study should offer guidelines to fire departments and insurors to lessen the likelihood that properties would be overvalued. or that fraudulent claims were paid due to a lack of appropriate cooperation between public investigator and insurors

One state has recently introduced legislation barring mortgage holders from requiring insurance premiums covering the full mortgage amount to the extent that the mortgage covers property value. This provision would tend to eliminate this source of over insurance

Insurors in high risk areas should be encouraged to use the Two-Part Insurance Application Form that industry agents developed to discourage arson-for-profit.

Insurors, underwriters, adjustors and their counterparts should develop better working procedures that deter or detect arson fraud schemes. Examples of such arrangements should be gathered and then disseminated through such channels as the Arson Resource Center, the Insurance Committee for Arson Control, and the Insurance Crime Prevention Institute.

Local and state investigators should be familiar with document search techniques so that they can identify insurors through public records.

Case Feedback

Need Statement:

Timely feedback to fire departments about cases in their areas.

Discussion:

It was widely acknowledged among participating fire officers and investigators that prompt feedback about case progress and disposition helps to maintain fire department interest in arson. This has a synergistic effect in that it can lead to better information flow between fire service personnel and fire investigators.

The primary requirements for such a feedback mechanism are that they be quick, simple, and inexpensive. As a courtesy, investigators should notify fire departmentsof thedisposition and status of cases in which they have an interest.

This is sound in concept but, as some investigators interviewed pointed out, difficult in practice due to the time requirement it imposes. Others expressed a concern that they might compromise an ongoing investigation or violate privacy standards.

■ Arson Control Expert Concurrence

Agree 23   Disagree 1

Solution Options:

One option that has proven effective in several settings involves state fire marshals announcing case clearances and convictions at meetings of the county firefighter's and county fire chiefs associations. This allowed the fire marshal to credit individuals and departments in a public setting but without revealing sensitive information. Another alternative involves investigators recording a videotaped briefing to update firefighters about the current status of cases. Here again, investigators need not reveal sensitive information or compromise ongoing investigations. Recorded briefings have certain advantages. They can reinforce training points, offer the further safeguard of being reviewed and edited before release, and can be replayed at the convenience of the user.

Inter-Agency Communication

Need Statement:

Improved communication among elements of the arson control system is needed to enhance cooperation.

Discussion:

Fire officers almost without exceptionbelieved that prosecutors failed to pursue cases vigorously enough to obtain convictions and appropriate punishments. Fire officers and firefighters said this dimmed their interest in going the extra mile to detect arson and preserve evidence. It is hard to see how it can be expected for volunteer firefighters to take the extra time and effort to detect arson, preserve the scene, and delay their return to their beds, jobs, or families if the rest of the system does not recognize their efforts, reward them (if by no other way than praise), and reinforce these efforts by feeding back information on:

- case status;
- case disposition.

For example, fire investigators often find it difficult to convince volunteer firefighters that plea bargaining is a necessary part of the criminal justice system. And in many instances, fire investigators disagree with the way a prosecutor handles a case. A prosecutor's decision to allow reduced charges or sentences may not be understood by firefighters or investigators. This in turn can harm their morale and undermine arson control efforts.

■ Arson Control Expert Concurrence

Agree 23   Disagree 0   No Comment 1

Solution Options:

Produce a simple guide that state fire marshals or state fire training systems can use to increase awareness among all concerned parties. Such a guide would outline the workings of the state's criminal justice system. The guide might explain, for example, what factors play a major role in a prosecutor's decision to pursue a case, why some cases are declined for prosecution, and why other cases are plea bargained. It might also suggest ways in which local fire organizations could monitor the disposition of arson cases and give appropriate feedback to prosecutors.

Combat Public Apathy Toward Arson

Need Statement:

Public apathy toward arson and arsonists hampers arson control efforts in rural areas.

Discussion:

Fire service personnel repeatedly made reference to public apathy as a major hindrance to their arson control efforts. The principal complaint was that people in rural areas see arson as a minor crime and one that will not strike them. Indeed, some forms of arson may not be regarded as a crime at all. For example: in rural areas of the South it has been common practice for generations to bum woods and brush to clear land and to control insect and snake populations.

One prosecutor pointed out that juries are unlikely to mete out stiff punishment for an offense they do not regard as a serious one. He noted that "ripping-off" an insurance company (by arson) is regarded in much the same way as cheating on income taxes: publicly condemned, but privately tolerated.

■ Arson Control Expert Concurrence

Agree 23   Disagree 1

Solution Options:

At least one insurance company has undertaken a media campaign to publicize the cost of arson to the public-that 25 cents of every property insurance premium dollar is lost to arson . Similar campaigns have been undertaken by public or quasi-public organizations (notably fire prevention cooperatives) with good effect.

Oklahoma has instituted an arson hotline with an "800" telephone number. Funded with insurance company money, and run by the state Arson Advisory Council, it pays $50 and up for tips. The Council prints and distributes" business cards" containing hotline information in addition to the usual promotional brochures, newsletters, and radio ads.

Public Education/Awareness

Need Statement:

Increasing public awareness of the cost of arson to the community and of its status as a serious criminal offense.

Discussion:

Each of the four work groups saw merit in educating the public and combatting public apathy about arson. However, none felt that they had the resources to produce their own public information/education materials. They stated a need for material that could be used directly or easily adapted to local needs. Benefits that could be derived are:

> Direct public support for arson control efforts (e.g., prosecution);
> A decrease in the number of set fires;
> Increased public willingness to provide information through tip lines; and,
> Increased level of support from other agencies and officials (e.g., from law enforcement, elected officials).

■ Arson Control Expert Concurrence

Agree 24   Disagree 0

Solution Options:

Work with community volunteer programs to ensure that arson is understood to be a rural problem as well as an urban one. The Southeast Oklahoma Fire Prevention Association, a fire prevention cooperative formed in 1978, uses public education and information about arson as key elements of its program. Most successful has been a 1982 public information film produced in an area with a high rate of incendiary fires. Using local funds and with local citizens as actors, the film has greatly increased public awareness of and support for arson control measures. The effort also includes a "Learn Not To Bum" program for juveniles through the school system. This resulted in a 33% reduction in incendiary fires in its first three years of operation.

The Florida Fire Marshal's office makes a special effort to publicize the stiff penalties for arson under the state's arson law. Publicity is also given to prosecutions and convictions.

The Klamath County, Oregon Arson Strike Force has had good cooperation from media in publicizing its efforts. Perhaps because of this, few trials result from the Strike Force program. The well-publicized efforts the Strike Force may increase willingness of defendants to plead guilty to a lesser charge due to a perceived likelihood of conviction.

The Indiana Fire Marshal's office operates an arson hotline using a toll-free phone number. The phone number, together with information on the program, is regularly broadcast on radio and TV stations around the state as a public service announcement.

Neighborhood Watch programs might be approached to assist in anti-arson and arson awareness activities.

www.ingramcontent.com/pod-product-compliance
Lightning Source LLC
Chambersburg PA
CBHW081413170526
45166CB00010B/3330